Berwyn The Alchemical Wizard With Beryllium Magic

Element 4 – Magical Elementals Chemistry Fantasy

By Sybrina Durant

Berwyn The Alchemical Wizard With Beryllium Magic

Element 4 – Magical Elementals Chemistry Fantasy

Paperback ISBN 978-1-942740-73-5

BISAC Codes:
JUV036000 - JUVENILE FICTION / Science & Technology
JUV037000 - JUVENILE FICTION / Science Fiction, Fantasy, Magic
JUV002270 JUVENILE FICTION / Fantasy / Dragons, Unicorns & Mythical Creatures

League City, Texas, United States of America

Contact Sybrina@sybrina.com.

No Metal
Berwyn
Berwyn The Wizard With
The Berylliam Staff
Be
No Magic

Chapter 1: The Spark of Beryllium

High in the Crystal Spire, where polished quartz walls caught the first light of dawn and split it into rainbows, Berwyn the Alchemical Wizard hummed a tune she'd once danced to at the Festival of Unbound Atoms. Her hair, a cascade of spun sunlight, swayed with every step as she circled a floating beryllium prism that pulsed with soft green luminescence. Large, curious eyes—the color of amber held up to a flame—studied the prism's facets, noting how each one bent the light into a different frequency. She wore a deep violet robe embroidered with golden spirals and ancient alchemical symbols, its hem brushing the crystalline floor. Matching gloves, stitched with threads of beryllium covered her fingers, and her wand—tipped with a shard of pure beryllium—rested in a loop at her belt.

"You're going to wear a groove in the floor, Berwyn," came a cheerful voice from the doorway. Kaelen, her oldest friend, leaned against the frame with a tool belt slung low over her leather trousers. She had no magic, but her fingers could coax life into the most stubborn old-world steam engines and new-world arcano-circuit boards alike. A pair of brass goggles perched on her forehead, and a multi-spanner glittered in her hand. "The prism won't reveal its secrets any faster if you pace."

Berwyn smiled without looking away. "Beryllium is patient, but I am not. Its atomic bonds are so light, so strong—imagine if I could weave that strength into a shield that covers an entire city. The Rust Alchemist's decay would shatter against it like glass."

Kaelen's expression sobered. The name hung in the air like a tarnish. For weeks, reports had trickled into Lumina—farmland turning brittle, iron bridges crumbling into orange dust, entire forests of ancient trees snapping under their own weight. The Rust Alchemist, a rogue elementalist who corrupted the bonds between atoms, was spreading his blight across MarBryn.

A soft chittering drew their attention. Zephyr, a foot-long sky-drake with scales that shimmered from copper to emerald, fluttered down from the rafters. His tiny wings hummed with residual static, and his voice was a crackle of lightning in miniature. "The council sent a messenger. A grumpy one. He's waiting in the atrium and says it's urgent."

Berwyn caught the prism with a flick of her fingers, the beryllium humming against her palm. She felt the familiar tingle of her power—the ability to sense and manipulate the invisible threads that held matter together. With a thought, she could lighten a boulder or reinforce a crumbling wall. Her force fields, crafted from woven atomic bonds, had turned aside arrows and even a rampaging stone golem. But the Rust Alchemist's magic attacked those very bonds, unraveling them like a knitter pulling a loose thread.

She strode to the atrium, Kaelen and Zephyr flanking her. The messenger, a wiry man in the cobalt robes of the Elemental Council, stood beside a flickering holographic map projected from a brass-and-crystal device—new-world technology powered by captured lightning. The map showed all ofMarBryn's provinces, and several were already bleeding rust-colored veins.

"Alchemical Wizard Berwyn," the messenger said, bowing hastily. "The Council requests your aid. The Rust Alchemist has seized the Sunken Laboratory beneath the Glass Desert. He's using old-world alchemical engines to amplify his corruption. If he succeeds, the decay will spread to the capital by the next full moon."

Kaelen whistled. "The Sunken Laboratory? That place was sealed centuries ago. It's full of unstable elementals and forgotten experiments."

"Which is exactly why he chose it," Berwyn murmured. She traced a finger over the map, her beryllium-tipped wand now in her other hand. "He's not just spreading decay—he's building something. Beryllium is used in alloys for precision instruments. If he's hoarding it, or corrupting its bonds, he could create a weapon that doesn't just rust metal but unravels all matter."

Zephyr buzzed nervously, a tiny spark leaping from his snout. "Then we stop him! I can scout ahead, and Kaelen's gadgets can disable traps."

"And I'll shield us," Berwyn said, her voice calm but edged with steel. The quiet confidence that had always been her hallmark settled over her like a mantle. She was a strategist, and already her mind was charting a path. "We leave at noon. Kaelen, bring your portable forge and the new resonance disruptor you've been tinkering with. Zephyr, I'll need you to carry a message to the Wind-Sisters at the Aerie—they might lend us a sky-skiff."

The messenger looked relieved. "The Council will provide provisions and a map of the old tunnels. But be warned: the Rust Alchemist has twisted the laboratory's guardians. You'll face things that were never meant to wake."

Berwyn nodded. She had faced such things before. Her empathetic nature reached out, sensing the messenger's genuine fear, and she placed a gloved hand on his shoulder. "We will return with his schemes in ruins."

As the messenger departed, Kaelen hefted her tool belt. "So, the Glass Desert. It's a three-day journey even by sky-skiff, and the desert is full of silica-wraiths that can turn a person to glass with a touch."

"Then I'll need to perfect my beryllium shield," Berwyn said. She lifted her wand, and a shimmering, nearly invisible dome flickered around them for an instant—a force field woven from strengthened atomic bonds. It hummed with a high, clear note, like a struck crystal goblet. "Beryllium is light but incredibly rigid. If I can tune the field to resonate at the exact frequency of the decay magic, it might neutralize it."

Zephyr landed on her shoulder, his scales warm. "And if it doesn't?"

Berwyn's amber eyes met his, and a hint of her playful spirit returned. "Then we improvise. I didn't become a master alchemist by following recipes exactly." She twirled her wand, and a cascade of golden sparks—alchemical motes that smelled of ozone and summer rain—swirled around her. "Now, let's pack. The Rust Alchemist has had his fun. It's time to show him what happens when you try to break the bonds that hold the world together."

With that, the three companions moved into action, the Crystal Spire humming with the promise of an adventure that would test the very limits of beryllium's magic and the bonds of friendship. Outside, the sun climbed higher over MarBryn, unaware that a wizard, a tinkerer, and a tiny dragon were about to challenge a darkness that threatened to unmake everything.

Chapter 2: The Glass Desert's Breath

The Aerie of the Wind-Sisters perched atop the jagged peaks of the Skyreach Mountains, a cluster of floating platforms tethered by shimmering energy cables to the stone below. Berwyn, Kaelen, and Zephyr arrived via a creaking cable-car that ran on old-world counterweights and new-world magnetic levitation—a testament to MarBryn's blend of technologies. As they stepped onto the main platform, the wind howled, carrying the scent of ozone and distant lightning.

Three women in flowing silver robes descended from a higher platform, their feet barely touching the ground. The Wind-Sisters, masters of air currents and weather magic, regarded the travelers with serene smiles.

"Berwyn of the Beryllium Wand," the eldest sister, Aella, greeted. "Zephyr's message reached us on the eastern gales. You seek a sky-skiff to cross the Glass Desert."

"Yes," Berwyn replied, her voice steady despite the altitude. "The Rust Alchemist must be stopped before his decay spreads further. We need speed and maneuverability."

Aella nodded, gesturing to a sleek vessel moored at the edge. The sky-skiff was a marvel: a hull of lightweight beryllium-aluminum alloy, old-world in its aerodynamic design but fitted with arcane sails that shimmered with captured wind elementals. A small arcano-turbine hummed at the stern, ready to convert ambient magical energy into thrust.

"The Silver Zephyr," Aella said. "She'll carry you swiftly. But beware—the Glass Desert is not merely sand and heat. Silica-wraiths roam the dunes, born from ancient alchemical spills. They can transmute flesh to glass with a touch."

Kaelen ran a hand along the hull, her engineer's mind already cataloging the craftsmanship. "This alloy—it's incredibly light. Beryllium, yes? I've only read about it in old texts. The tensile strength is off the charts."

Berwyn smiled. "The same element that fuels my magic. It's rare and difficult to work, but the Wind-Sisters have mastered its forging. I'll reinforce the hull with a beryllium force field once we're airborne. That should give us an edge against the wraiths."

Zephyr fluttered his wings excitedly. "I'll ride the bow! I can sense air currents and spot danger from afar."

With supplies loaded—water skins, dried fruit, Kaelen's portable forge, and a resonance disruptor prototype—the companions boarded. Berwyn took the helm, her wand glowing as she channeled a thin, invisible layer of strengthened atomic bonds around the skiff. The force field shimmered briefly, then settled into transparency.

"Hold tight," Berwyn said, and with a thought, she lightened the vessel's mass. The skiff rose effortlessly, sails snapping full as Aella summoned a favorable wind. They soared southward, leaving the mountains behind.

For hours, the landscape unfolded below: patchwork farmlands, then the shimmering expanse of the Glass Desert. It was a sea of crystalline dunes that refracted sunlight into blinding prisms. The heat was intense, but Berwyn's shield also insulated them from the worst.

Kaelen tinkered with her disruptor, a device that emitted counter-frequency vibrations. "If the Rust Alchemist's magic works by breaking atomic bonds, maybe I can tune this to stabilize them instead. It's experimental, but it might help against his decay."

"A sound theory," Berwyn said, her eyes scanning the horizon. "Beryllium's strength comes from its electron configuration—it forms strong covalent bonds. If I can amplify that resonance, we might be able to reverse his corruption in small areas."

Zephyr suddenly shrieked, a crackle of electricity arcing from his snout. "Wraiths! Three of them, rising from the dune ahead!"

Berwyn saw them: shimmering, vaguely humanoid figures made entirely of living glass. Their forms shifted, refracting light into deadly beams. One raised an arm, and a ray of concentrated sunlight shot toward the skiff.

Berwyn reacted instantly, extending her wand. The beryllium force field hardened, and the beam splintered into harmless rainbows. "Kaelen, take the helm! Zephyr, distract them with lightning!"

Kaelen grabbed the tiller while Berwyn stood, planting her feet. She focused on the wraiths, sensing the silica bonds that held their glass bodies together. They were unstable, crackling with trapped energy. With a precise gesture, she sent a pulse of beryllium magic into the nearest wraith—not to break its bonds, but to reinforce them, locking its atoms into a rigid lattice.

The wraith froze mid-lunge, its glass body becoming perfectly rigid. It toppled, shattering into harmless shards.

"One down!" Zephyr cheered, diving at the second wraith. He released a bolt of lightning, but the creature absorbed the energy, glowing brighter. It retaliated with a wave of transmuting sand.

Berwyn threw up a second shield, but the sand clung to it, trying to crystallize. She grimaced, feeling the strain. "They're adapting! Kaelen, the disruptor—try it now!"

Kaelen aimed the device, its brass coils humming. A low-frequency vibration pulsed outward, and the second wraith's form wavered, its bonds destabilizing. It lost cohesion, melting into a puddle of inert silica.

The third wraith, however, had circled behind. It lunged at the skiff's stern, its glassy claws raking the hull. The force field held, but Berwyn felt a cold seep through—the wraith's touch was trying to transmute the metal.

"Enough!" Berwyn slammed her wand against the deck. A wave of pure beryllium energy radiated outward, not a shield but a controlled burst of atomic strengthening. The skiff's hull became momentarily impervious, and the wraith's claws shattered on impact. The creature recoiled, then fled back into the dunes.

Breathing heavily, Berwyn lowered her wand. "Is everyone all right?"

"Just a few scratches on the paint," Kaelen said, patting the hull. "Your magic held. That last trick—you reinforced the entire ship's atomic structure?"

Berwyn nodded, wiping sweat from her brow. "Beryllium's lightness combined with its rigidity. I can't do it for long, but it's effective in bursts. The wraiths are a problem, though. They're drawn to magical energy. We'll have to be more careful."

Zephyr landed on her shoulder, his scales dimmed. "The Sunken Laboratory is still half a day's flight. If the Rust Alchemist has more guardians like these..."

"Then we'll face them together," Berwyn said firmly. She looked at her companions—Kaelen's steady hands on the tiller, Zephyr's loyal spark—and felt a surge of gratitude. Her empathetic nature sensed their resolve, their trust. "We have the tools, the knowledge, and the will. Now, let's find a place to rest and plan. The real battle is yet to come."

As the Silver Zephyr sailed deeper into the glittering desert, the sun began its descent, painting the glass dunes in shades of fire and blood. In the distance, a dark smudge on the horizon marked the location of the Sunken Laboratory—a place where old-world alchemy had once thrived, and where a new evil now festered. Berwyn's beryllium wand glowed softly, ready for the trials ahead.

Chapter 3: The Sentinel of Shifting Sands

The Silver Zephyr descended through a layer of heat shimmer that warped the horizon into liquid waves. Below, the Glass Desert gave way to a vast crater—the Sunken Laboratory, a relic of MarBryn's most ambitious and dangerous era. Spires of twisted metal and crystallized alchemical residue jutted from the sand like the bones of some ancient beast. At the crater's center, a massive dome of cracked obsidian reflected the dying sun in fractured patterns.

Berwyn guided the skiff toward a flat outcropping of fused glass that served as a natural landing pad. As they touched down, she released the beryllium-enhanced lightness, and the vessel settled with a gentle crunch. The force field shimmered and faded, conserving her energy.

"Look at this place," Kaelen breathed, hopping out with a portable scanner—a new-world device that hummed as it analyzed ambient magic. "Residual alchemical signatures are off the charts. There's beryllium, iron, copper, and something else... something unstable."

Zephyr fluttered to a high perch, his scales crackling. "I don't like it. The air tastes like rust and lightning."

Berwyn stepped onto the glassy ground, her boots clicking. She extended her senses, feeling the atomic bonds around her. The very sand vibrated with tension, as if the desert itself was holding its breath. "The Rust Alchemist has been here. I can feel the decay—it's like a low hum, eating away at the edges of things. But there's something else... a presence. Ancient. Watchful."

They moved cautiously toward the dome's entrance, a yawning archway flanked by two statues of winged serpents. The statues were carved from obsidian, their eyes set with dull gems. As Berwyn approached, the gems flickered to life—a deep, angry red.

"Intruders," a voice boomed, not from the statues but from the ground itself. The sand before the archway erupted, coalescing into a towering figure. It was humanoid, easily fifteen feet tall, its body composed of compacted silica and crystalline shards. Veins of rust-colored corruption pulsed through its form like diseased arteries. Its eyes were twin pools of molten glass, and when it spoke, its voice was the grinding of tectonic plates.

"I am Terron, Sentinel of the Sunken Laboratory. Once, I guarded the secrets within. Now, I serve the Rust Alchemist. You will not pass."

Kaelen raised her resonance disruptor. "It's a silica elemental! The corruption is overriding its original programming."

Berwyn stepped forward, her wand glowing. "Terron, listen to me. You were created to protect, not to destroy. The Rust Alchemist has poisoned your core. Let me help you."

The sentinel laughed, a sound like shattering glass. "Help? The Rust Alchemist gave me purpose. The old masters abandoned me. He filled the void with his power. Now, I am complete."

It raised a massive arm, and the sand around them erupted into razor-sharp glass shards. Berwyn reacted instantly, conjuring a beryllium force field that deflected the projectiles. The shards shattered against the invisible barrier, raining harmless dust.

"Zephyr, aim for its joints!" Berwyn commanded. "Kaelen, tune the disruptor to the frequency of silica bonds—try to destabilize the corruption, not the elemental itself!"

Zephyr dove, releasing a concentrated bolt of lightning at the sentinel's knee. The electricity arced through the crystalline structure, causing it to stumble. Kaelen frantically adjusted dials on her device, muttering calculations under her breath. "Silica's bond frequency is around 1100 inverse centimeters in the infrared... if I can match that and overlay a counter-resonance..."

The sentinel roared, slamming its fist into the ground. A shockwave rippled outward, turning the sand beneath Berwyn's feet into a sinkhole of liquid glass. She lightened her own mass with a thought, leaping backward just in time. Her wand traced a spiral, and she sent a pulse of pure beryllium energy into the sentinel's chest—not to break, but to stabilize.

For a moment, the rust-colored veins flickered, and the sentinel froze. Its molten eyes dimmed, then brightened again with a different light—a clear, crystalline blue.

"I... remember," Terron rumbled, its voice no longer grinding but resonant, like a deep bell. "The old masters... they gave me a purpose. To guard knowledge. To preserve, not to destroy. But the Rust Alchemist... he twisted that purpose. He made me his weapon."

Berwyn lowered her wand, her force field still shimmering. "The corruption is still there, but I've reinforced your core bonds. Fight it, Terron. You are stronger than his decay."

The sentinel trembled, its massive form cracking as the rust veins pushed back. "It... burns. He will not release me so easily."

Kaelen stepped forward, her disruptor now humming at full power. "I've got the frequency! Berwyn, on my signal, amplify the beryllium resonance. I'll hit it with a destabilizing pulse—together, we can purge the corruption!"

Berwyn nodded, placing a gloved hand on the sentinel's leg. She felt the battle within Terron's atomic structure: the Rust Alchemist's decay magic was like a cancer, breaking bonds and replacing them with brittle, oxidized chains. Her beryllium magic could strengthen, but she needed to target the corruption precisely.

"Now!" Kaelen shouted, and the disruptor emitted a high-pitched whine that made the air vibrate.

Berwyn channeled her power through her wand and into Terron's core. She visualized the beryllium atoms—light, strong, with a unique electron configuration that formed rigid covalent bonds. She wove those bonds into the sentinel's silica matrix, reinforcing the healthy structure while pushing against the rust. The corruption fought back, but with Kaelen's disruptor shaking its hold, it began to crumble.

Terron let out a cry that was part agony, part relief. The rust veins shattered, turning to dust that blew away on the desert wind. The sentinel's body, now pure crystalline silica, gleamed in the fading sunlight. Its eyes glowed a steady, peaceful blue.

"I am free," Terron said, its voice now gentle. "You have my gratitude, Alchemical Wizard. I have been a prisoner in my own body for too long."

Berwyn smiled, lowering her wand. "You were never truly gone, Terron. Your core purpose remained, buried beneath the corruption. Will you help us stop the Rust Alchemist?"

The sentinel bowed, a surprisingly graceful motion for a being of stone and glass. "I know the laboratory's secrets—its hidden passages, its defenses. The Rust Alchemist has made his lair in the old transmutation chamber. I will guide you, and I will fight beside you."

Zephyr landed on Berwyn's shoulder, eyeing the towering elemental. "Well, we've certainly upgraded our party. A fifteen-foot silica sentinel. That'll make an impression."

Kaelen stowed her disruptor, grinning. "And I finally got that thing to work. Told you it wasn't just a fancy paperweight."

As the group gathered at the archway, Terron raised a hand, and the obsidian doors groaned open. Beyond lay darkness, and the faint, acrid smell of rust. Somewhere deep within, the Rust Alchemist was waiting, his schemes nearing completion.

Berwyn stepped forward, her beryllium wand held high, its tip igniting with a soft green flame. "Let's finish this. Together."

The companions—a wizard, a tinkerer, a sky-drake, and a sentinel—entered the Sunken Laboratory, ready to face the decay that threatened all ofMarBryn.

Chapter 4: The Transmutation Chamber's Secret

The corridor beyond the obsidian doors sloped downward, its walls lined with ancient tiles that had once been white but now bore the creeping orange stain of rust. Old-world glow-globes, powered by residual alchemical energy, flickered feebly overhead, casting long shadows that danced like specters. Berwyn led the way, her beryllium wand held aloft, its green flame pushing back the darkness. Kaelen walked beside her, scanner humming, while Zephyr perched on Terron's broad shoulder, the sentinel's crystalline body refracting the faint light into prismatic patterns.

"The air is thick with decay," Terron rumbled, his voice echoing softly. "The Rust Alchemist has been here for weeks, twisting the old machinery to his purposes. The laboratory was built to transmute base metals into gold, but its true purpose was far more ambitious—the manipulation of atomic bonds on a grand scale. He is using that power to spread his corruption, but I sense he is building something more."

"What do you mean?" Berwyn asked, pausing at a junction where three tunnels branched. She extended her senses, feeling the atomic vibrations. The left tunnel hummed with a low, steady rhythm—the heartbeat of dormant machines. The right tunnel was silent, dead. The center one pulsed erratically, a chaotic jangle of unstable bonds.

Terron pointed a massive finger toward the center. "That leads to the transmutation chamber. But it is guarded. The Rust Alchemist has twisted the old security constructs—they are no longer mere golems. He has infused them with decay magic, making them unpredictable."

Kaelen adjusted her goggles, peering into the darkness. "I'm reading massive energy fluctuations. If we go that way, we'll need to be ready for anything. My disruptor is still charged, but I'll need time to recalibrate for whatever he's cooked up."

Berwyn considered their options. Her empathetic nature reached out, sensing the lingering pain in the walls—the laboratory itself seemed to mourn its corruption. "We can't afford to waste time. If he completes whatever he's building, the decay could spread beyond the desert. Terron, you said you know hidden passages. Is there a way to bypass the main defenses?"

The sentinel nodded slowly, his crystalline joints grinding softly. "There is a maintenance shaft. It was used by the old masters to service the arcane conduits. It leads directly to the observation deck above the transmutation chamber. But it is narrow—I will not fit."

"Then you'll guard our exit," Berwyn decided. "Kaelen, Zephyr, and I will take the shaft. We'll assess the situation and, if possible, disrupt his work before engaging directly."

Zephyr fluttered his wings. "I can squeeze through easily. I'll scout ahead."

Terron placed a hand against the left wall, and a section of tile slid aside, revealing a dark, cramped passage lined with dusty pipes and flickering runes. "Follow this path. It will bring you to a grate overlooking the chamber. Be silent—the Rust Alchemist's senses are attuned to disturbances in atomic bonds. Your beryllium magic will be a beacon if you are not careful."

Berwyn nodded, dimming her wand's flame to a faint glow. She and Kaelen squeezed into the shaft, Zephyr flitting ahead. The passage was tight, and the air was stale, heavy with the smell of ozone and old metal. They crawled for what felt like an eternity, until a faint orange light appeared ahead.

Zephyr returned, his voice a whisper of static. "The chamber is huge. There's a massive machine at the center—like a giant forge, but it's made of glass and copper. And there are cages. I saw creatures inside, glowing with elemental energy. The Rust Alchemist is siphoning their power into something."

Berwyn's heart clenched. "He's using living elementals as fuel. We have to stop him."

They reached the grate, which overlooked a vast circular chamber. The transmutation chamber was a cathedral of old-world science and new-world magic. At its center stood a towering apparatus: a spherical crucible of reinforced crystal, surrounded by copper coils that crackled with arcane lightning. Pipes fed into it from all sides, carrying streams of rust-colored energy. Along the walls, a dozen cages held captive elementals—a fire sprite, a water naiad, an earth gnome, and others, all weakened, their light flickering.

And there, standing before the crucible with his back to them, was the Rust Alchemist. He was tall and gaunt, draped in robes the color of dried blood. His hands, skeletal and tipped with iron claws, gestured over a control panel that blended old-world levers with new-world holographic displays. A mask of corroded copper hid his face, but his voice echoed through the chamber, cold and methodical.

"Soon, the Beryllium Heart will be complete. With it, I will unmake the bonds that hold matter together—not just metal, but stone, flesh, and bone. MarBryn will crumble, and from the rust, a new order will rise."

Kaelen's breath caught. "Beryllium Heart? He's building a weapon that uses beryllium's properties in reverse? That's... that's insane. Beryllium is a stabilizer, not a destroyer."

Berwyn's mind raced. Beryllium's strength came from its electron configuration—it formed strong covalent bonds and was used in alloys to add rigidity and lightness. But if one were to invert that resonance, to force beryllium atoms to repel instead of bond, it could theoretically unravel any material that contained even trace amounts. And beryllium was present in countless minerals across MarBryn.

"He's not just corrupting," she whispered. "He's planning a chain reaction. If that Heart activates, it will spread through the earth itself, breaking apart everything."

Zephyr's scales bristled. "We have to smash it!"

"Not yet," Berwyn said, her strategic mind taking over. "If we destroy it carelessly, the unstable energy could detonate. We need to free the elementals first—their power is feeding the weapon. Kaelen, can your disruptor interfere with the energy flow without causing a cascade?"

Kaelen studied the apparatus, her fingers twitching. "Maybe. Those copper coils are conducting the siphoned elemental energy into the crucible. If I can tune the disruptor to the exact frequency of the elemental bonds, I might be able to create a feedback loop that shuts it down safely. But I'll need to get closer, and it'll take time."

Berwyn nodded. "Then we create a diversion. Zephyr, you'll cause a lightning storm near the ceiling—make him think the laboratory's old wards are activating. I'll use my force fields to protect the elementals while Kaelen works. Once the Heart is destabilized, we confront him directly."

They moved into position. Zephyr slipped through the grate and soared upward, his tiny body crackling with building charge. Berwyn focused on the cages, mapping the atomic bonds of the bars. They were iron, heavily corroded, but she could reinforce them temporarily to prevent collapse while she worked on the locks.

Zephyr unleashed a dazzling display, bolts of lightning arcing across the ceiling. The Rust Alchemist spun, his mask reflecting the flashes. "What—? The old wards? Impossible! They were deactivated centuries ago!"

Seizing the moment, Berwyn dropped from the grate, her wand already weaving a shimmering dome around the cages. The force field, woven from strengthened beryllium bonds, encased the elementals, protecting them from any stray energy. She landed silently, then sprinted toward the cages.

Kaelen slid down a pipe, disruptor in hand, and crept toward the copper coils. She knelt, opening a panel and beginning her adjustments, muttering under her breath. "Frequency modulation... phase inversion... come on, come on..."

The Rust Alchemist spotted Berwyn. "You! The beryllium witch. I sensed your magic the moment you entered my domain. Did you think you could sneak past me?" He raised a hand, and a wave of rust-colored energy surged toward her.

Berwyn threw up a second shield, the force field absorbing the decay magic with a shriek of stressed atoms. She gritted her teeth, feeling the bonds strain. "Your corruption ends here, Alchemist. I won't let you use beryllium to destroy what it was meant to protect."

He laughed, a hollow sound. "Beryllium is a tool, like any other. Its strength can build, but it can also break. You cling to your narrow view of magic, but I see the bigger picture. From decay comes renewal. MarBryn is stagnant—I will cleanse it."

"By killing innocent elementals?" Berwyn shot back, nodding toward the cages. "That's not renewal. That's murder."

She reached the first cage, touching the lock. Her magic flowed into the iron, reinforcing its bonds just enough to make it rigid, then she shattered it with a precise pulse. The fire sprite inside flickered weakly. "Go," she whispered. "Get to safety."

The sprite flared and zipped away. Berwyn moved to the next cage, and the next, freeing the captives one by one. The Rust Alchemist howled in fury, sending another wave of decay, but Berwyn's shield held. Each freed elemental weakened the Heart's power supply, and the crucible's glow began to dim.

"Now, Kaelen!" Berwyn shouted.

Kaelen slammed the disruptor's activation switch. A high-frequency pulse rippled through the copper coils, and the energy flow reversed. The crucible shuddered, cracks spiderwebbing across its crystal surface. The Rust Alchemist stumbled, his control panel sparking.

"No! My Heart!" He lunged toward the machine, but Berwyn was faster. She lightened her body, leaping between him and the crucible, her wand blazing. "It's over. Stand down."

For a moment, the chamber was silent save for the crackling of dying energy. The Rust Alchemist's mask tilted, and Berwyn felt a wave of pure hatred. "This is not the end, wizard. You've only delayed the inevitable."

He slammed his iron claws together, and a cloud of rust particles exploded outward, obscuring everything. When it cleared, he was gone, a hidden passage grinding shut behind him.

Kaelen rushed to Berwyn's side. "Did we do it? Is the Heart disabled?"

Berwyn examined the cracked crucible. The energy was dissipating, but the core—a shard of corrupted beryllium—still pulsed faintly. "He'll be back. But for now, we've saved the elementals and stopped his immediate plan. Let's secure this chamber and find where he fled."

Zephyr landed, exhausted but triumphant. "That was close. But we're not done, are we?"

"No," Berwyn said, her amber eyes hardening. "We're just getting started."

Chapter 5: The Foundry of Forgotten Alloys

The transmutation chamber fell silent save for the faint hiss of escaping steam and the labored breathing of the freed elementals. Berwyn knelt beside the cracked crucible, her wand hovering over the corrupted beryllium shard at its core. The shard pulsed with a sickly orange light, its atomic bonds twisted into a perversion of their natural state. She could feel the decay radiating from it like heat from a dying star.

"This shard is still dangerous," she said, her voice tight with concentration. "If left unchecked, it could seed new corruption. I need to contain it."

Kaelen rummaged through her tool belt and produced a small lead-lined box, originally designed to store radioactive alchemical samples. "Will this work? It's shielded against most energy emissions."

Berwyn nodded, and with a careful application of her beryllium magic, she encased the shard in a temporary force field, lightening it so she could levitate it into the box. The lid clicked shut, and the orange glow vanished. "That will hold for now. But we need to find the Rust Alchemist. He escaped through a hidden passage. Terron might know where it leads."

Zephyr, still recovering from his lightning display, fluttered weakly onto Kaelen's shoulder.

"The elementals want to thank you. The fire sprite—her name is Embera—says she can guide us through the lower levels. She was held captive for weeks and learned the patrol routes of the corrupted constructs."

As if summoned, the fire sprite materialized before them, her form a dancing flame in the shape of a small woman. Her voice crackled like a campfire. "Alchemical Wizard, you saved us. I am in your debt. The Rust Alchemist has a secondary laboratory deeper down, in the Foundry of Forgotten Alloys. That's where he kept the worst of his experiments. I can take you there, but the path is treacherous."

Berwyn considered the offer. The sprite was weak, but her knowledge was invaluable. "We accept your guidance, Embera. But first, we must regroup with Terron. He's guarding the entrance. Kaelen, can you rig a signal to call him?"

Kaelen pulled out a flare device—a new-world gadget that emitted a specific arcano-frequency pulse. She aimed it at the ceiling and fired. A silent vibration rippled through the stone. "Terron will feel that through the silica in the walls. He'll find a way to us, even if it means digging."

They waited, and soon the chamber trembled as a section of wall crumbled inward. Terron's massive crystalline form squeezed through, his body shedding dust and small shards. "I sensed your victory, but also a retreat. The Rust Alchemist fled to the depths. I know the Foundry. It was where the old masters forged beryllium alloys and conducted their most dangerous transmutations. It is heavily shielded against magical detection, which is why I could not sense it before."

"Then that's where he's gone to ground," Berwyn said. "Embera will guide us through the patrol routes. Terron, can you clear the larger obstacles? We'll need your strength."

Terron inclined his head. "I will break whatever stands in our way."

The group descended through a spiraling staircase carved from obsidian, its steps worn smooth by centuries of use. Embera floated ahead, her flame illuminating ancient murals depicting alchemists bending atomic bonds. The air grew heavier, tinged with the metallic tang of old alloys and the faint, acrid scent of rust.

"The Foundry is just beyond this corridor," Embera whispered, her flame dimming. "But there's a guardian—a construct the Rust Alchemist twisted into a personal bodyguard. It's called the Ferroclast. It can reshape its metal body into weapons and absorb kinetic energy. I saw it crush a stone golem with a single blow."

Kaelen checked her disruptor. "Absorb kinetic energy? So hitting it makes it stronger. We'll need to use energy attacks or find a way to lock its atomic structure."

Berwyn's mind raced. "Beryllium can strengthen bonds, but it can also make them rigid to the point of brittleness. If I can force the Ferroclast's metal into a perfectly rigid lattice, it might shatter under its own weight. But I'll need to get close."

Zephyr puffed his chest. "I can distract it with lightning. That's energy, not kinetic. And Embera can use heat to warp its metal."

Terron stepped forward. "I will engage it directly. My silica body is not metal; it may not be able to absorb my strikes. And I am heavy enough to test its rigidity."

They advanced into the Foundry, a cavernous hall filled with massive crucibles, hydraulic presses, and arcane forges that still glowed with residual heat. At the far end, a figure of gleaming, rust-streaked steel stood motionless. The Ferroclast was humanoid but featureless, its limbs ending in blades and hammers that shifted and reformed with a liquid grace. Its eyes were twin furnaces of molten iron.

"Intruders detected," it intoned, its voice the screech of grinding gears. "Elimination protocol active."

It lunged, its arm transforming into a massive axe. Terron met it head-on, his crystalline fist colliding with the blade in a shower of sparks. The Ferroclast absorbed the kinetic energy, its body glowing brighter, but Terron's silica composition seemed to confuse it—the energy had no metal to transfer into.

"Now, Zephyr!" Berwyn shouted.

The sky-drake unleashed a bolt of lightning that arced into the construct's chest. The Ferroclast staggered, its glow flickering. Embera added a stream of concentrated flame, heating its joints to near-melting. The metal began to warp, but the construct adapted, sprouting cooling fins from its back.

"It's learning!" Kaelen warned. "Berwyn, do it now!"

Berwyn darted forward, her wand blazing with green light. She pressed her free hand against the Ferroclast's leg, ignoring the searing heat. She focused on the iron atoms, their bonds already strained by the thermal shock. With a surge of beryllium magic, she forced those bonds into a perfectly rigid, crystalline lattice—the same structure that made beryllium so strong but also brittle under extreme stress.

The Ferroclast froze. Its body, now locked in place, emitted a high-pitched whine as internal stresses built. Cracks spiderwebbed across its surface, and with a deafening crack, it shattered into a thousand fragments of inert steel.

Berwyn stumbled back, breathing hard. "That... took more out of me than I expected."

Terron caught her, his crystalline hand surprisingly gentle. "You did well. The way is clear."

Embera flared with pride. "The Rust Alchemist's secondary lab is through that door. But be careful—he'll have set traps."

Kaelen examined the shattered construct, pocketing a small sample of the alloy. "This metal is fascinating. It's an old-world memory alloy, but corrupted. I might be able to reverse-engineer a countermeasure if we face more like it."

They approached the heavy iron door at the far end of the Foundry. Berwyn could feel the Rust Alchemist's presence beyond, a cold knot of decay. She turned to her companions—Kaelen, ever resourceful; Zephyr, brave and loyal; Terron, a redeemed guardian; and Embera, a new ally forged in fire. Her empathetic nature swelled with gratitude and determination.

"Together, we've overcome every obstacle he's thrown at us," she said. "This ends now. Whatever he's planning in there, we'll stop it. Are you ready?"

A chorus of affirmatives answered her. Berwyn raised her wand, the beryllium tip glowing with renewed power. She pushed open the door, and the confrontation began.

Chapter 6: The Resonance of Ruin

The door swung open on silent hinges, revealing a chamber that was part laboratory, part mausoleum. Rows of old-world analytical engines—brass and iron monstrosities covered in dials and punched-card readers—lined the walls, their gears frozen mid-calculation. New-world arcano-crystals pulsed feebly beside them, wired into a central console that hummed with malevolent energy. At the room's heart stood a second crucible, smaller than the one above but far more refined, its crystalline surface etched with alchemical sigils that glowed a sickly orange.

The Rust Alchemist stood before it, his back still turned, his iron-clawed hands dancing over a holographic interface that projected atomic structures into the air. He was muttering, his voice a rasp of obsession. "Almost... the resonance is nearly perfect. Once I synchronize the decay frequency with the planetary beryllium veins, the chain reaction will be unstoppable."

Berwyn stepped forward, wand raised. "Step away from the console, Alchemist. Your scheme ends here."

He turned, the copper mask reflecting the orange light in twisted patterns. "Ah, the beryllium witch returns. And you've brought your menagerie. How quaint. Do you truly believe you can stop entropy itself? Decay is the natural state of all things. I am merely accelerating the inevitable."

"Decay is part of the cycle," Berwyn countered, her amber eyes blazing, "but so is renewal. You're not a force of nature—you're a madman imposing your will on a world that doesn't want it."

The Rust Alchemist laughed, a hollow, grating sound. "You speak of cycles? I was once like you—an alchemist who believed in balance. I studied beryllium for decades, marveling at its strength, its lightness. I used it to build, to stabilize, to heal. But then I saw the truth: matter is temporary. Bonds break. Everything returns to rust. The only true permanence is oblivion."

Kaelen, who had been scanning the console with her device, looked up sharply. "Wait—you're Doctor Aldric Voss? The alchemist who disappeared fifty years ago? You pioneered beryllium alloy synthesis! You were a hero!"

The Rust Alchemist—Voss—paused, his mask tilting. "Hero? I was a fool. An accident in this very foundry exposed me to a concentrated decay field of my own making. It unbound the atoms in my body, slowly. I should have died, but I learned to control the process, to spread it. Now I see that the only way to end suffering is to end the illusion of permanence. MarBryn will be the first to know true peace—the peace of dissolution."

Embera flared angrily. "You imprisoned my kind to fuel your madness! That is not peace—it's cruelty!"

Voss waved a dismissive claw. "The elementals are raw material, nothing more. Their energy is uniquely suited to propagate the decay frequency. Once I tap into the planetary beryllium network, the resonance will spread through the earth itself. Every mountain, every city, every living thing that contains even a trace of beryllium—which is most of them—will crumble."

Terron rumbled, stepping forward. "Then we will stop you here, as we stopped your Heart above."

"That was a prototype," Voss sneered. "This is the Resonator, and it is already active. Even now, its signal is reaching the deep beryllium deposits beneath the Glass Desert. In minutes, the chain reaction will begin." He pressed a final sequence on the console, and the crucible flared blindingly bright. A low, thrumming vibration filled the room, rattling the old machinery.

Berwyn felt it in her bones—a dissonant frequency that clawed at the atomic bonds around her. Her beryllium wand vibrated painfully. "He's not bluffing! We have to shut it down!"

"I'll disrupt the power flow!" Kaelen shouted, sprinting toward a junction box. "But I need cover!"

"Zephyr, Embera, with me!" Berwyn commanded. "Terron, smash the crucible if you can, but be careful—it might be shielded!"

The sentinel charged, his crystalline fists hammering against the Resonator's casing. Cracks appeared, but the device held, its sigils flaring defensively. The Rust Alchemist raised both hands, and tendrils of rust-colored energy lashed out—not at Terron, but at the old analytical engines. They groaned to life, their gears spinning, and from their innards crawled constructs of twisted metal and paper, animated by decay magic.

"Meet my Archivists," Voss hissed. "They are programmed to protect the Resonator at all costs."

The constructs scuttled forward on spindly legs, their punched-card bodies fluttering with malevolent intent. One spat a stream of corrosive ink at Zephyr, who dodged, his scales sizzling. Another swung a pendulum blade at Embera, who countered with a jet of flame, melting its brass joints.

Berwyn threw up a beryllium force field around Kaelen, who was frantically rewiring the junction box. "I've almost got it! The power source is a self-contained arcano-reactor—if I can reverse the polarity, it should feedback and shut down the Resonator!"

"Do it!" Berwyn turned to face Voss directly. "You were a healer once. There's still good in you. Let us help you."

Voss hesitated, his mask flickering with some buried emotion. "The good died with my body. This is all that remains." He lunged at Berwyn, claws extended.

She met him with her wand, a blade of pure beryllium energy forming at its tip—not a flame, but a concentrated edge of strengthened atomic bonds. They clashed, magic against decay, green light against orange. Berwyn felt the wrongness of his touch, the way his power tried to unweave her very being. But her beryllium magic held, reinforcing her own molecular structure.

"You can't win, girl," Voss snarled. "Even if you stop the Resonator, I have already set events in motion. The decay frequency has been transmitted to my followers across MarBryn. They will activate secondary resonators in every major city. You can't be everywhere at once."

Berwyn's heart sank, but she didn't relent. "Then we'll stop them one by one. Starting with you."

Zephyr dove from above, releasing a concentrated lightning strike that overloaded the Archivists' punch-card brains. They collapsed in heaps of sparking metal. Embera followed up with a firestorm that melted the remaining constructs into slag.

"Now, Kaelen!" Berwyn shouted.

Kaelen slammed the final connection. The arcano-reactor whined, its energy flow reversing. The Resonator's hum became a shriek, and the crucible shattered, spraying harmless crystal shards across the room. The orange glow died.

Voss screamed, staggering back. The decay tendrils retracted. "No! My work! You've ruined years of preparation!"

But even as he raged, his body began to destabilize. Without the Resonator's energy to sustain him, the decay that had kept him alive was now consuming him. He clutched his chest, his iron claws rusting before their eyes. "This... is not... the end..."

He collapsed, his form crumbling into a pile of inert dust and corroded metal. The copper mask clattered to the floor, empty.

The chamber fell silent. Berwyn lowered her wand, breathing hard. "Is everyone all right?"

Kaelen emerged from behind the junction box, soot-streaked but grinning. "I reversed the polarity. Classic trick. But Voss said there are more resonators. We have to warn the Elemental Council."

Embera flickered. "I can send word through the elemental network. Fire sprites can travel through heat vents and magma flows faster than any messenger. I'll alert the other enclaves."

Terron knelt, examining the dust that had been the Rust Alchemist. "He was once a great mind. It is tragic that his genius was twisted so."

Zephyr landed on Berwyn's shoulder. "So he's really gone? It feels... too easy."

Berwyn shook her head. "He said he transmitted the frequency. He may have been lying to distract us, but we can't take that chance. We need to get back to Lumina and report to the Council. If there are more resonators, we'll need help to find and destroy them."

She picked up the copper mask, turning it over. It was cold, inert. But as she held it, she felt a faint, lingering vibration—a residual decay signal, like a ghost in the metal. "This mask might hold a clue to his network. Kaelen, can you analyze it?"

Kaelen took the mask, her scanner humming. "It's got a unique arcano-signature. If I can trace it, we might be able to locate the other resonators before they activate. But I'll need better equipment—back at the Crystal Spire."

"Then that's our next destination," Berwyn said. "Embera, go with our thanks. Spread the word. Terron, will you accompany us? Your strength and knowledge of the old laboratories could be invaluable."

Terron nodded. "I will go where I am needed. The old masters built many hidden facilities. If Voss had followers, they may have occupied those sites."

Zephyr stretched his wings. "Let's get moving, then. The desert is still full of wraiths, and we've got a long flight back."

As they made their way out of the Foundry, Berwyn felt a mix of relief and urgency. The Rust Alchemist was defeated, but his legacy of decay lived on. MarBryn was still in danger, and she knew that the next phase of their journey would take them far beyond the Glass Desert—to the heart of the conspiracy that threatened to unmake their world.

The Silver Zephyr awaited, its sails gleaming in the moonlight that now filtered through the crater's rim. Berwyn looked back at the Sunken Laboratory one last time, then turned toward the future. The adventure was far from over.

Chapter 7: The Copper Mask's Secret

The Silver Zephyr soared over the moonlit dunes, its arcane sails billowing with a steady wind summoned by Zephyr's residual static. Berwyn sat at the helm, her amber eyes fixed on the horizon where the Skyreach Mountains rose like jagged teeth. The copper mask of the Rust Alchemist lay wrapped in a lead-lined cloth in her lap, its faint decay signal a constant, unsettling hum against her beryllium senses. Kaelen dozed in the passenger seat, her scanner clutched to her chest, while Terron stood motionless at the stern, his crystalline body refracting starlight into scattered rainbows.

By dawn, Lumina's Crystal Spire came into view, its quartz walls catching the first light and blazing like a beacon. Berwyn guided the skiff to the landing platform, where a delegation from the Elemental Council awaited—three robed figures, their faces grave. The eldest, a woman with silver-streaked hair and eyes like storm clouds, stepped forward.

"Alchemical Wizard Berwyn, we felt the disturbance in the desert's atomic bonds. The Rust Alchemist's presence has vanished. Is it done?"

Berwyn disembarked, her companions following. "The Rust Alchemist—Doctor Aldric Voss—is destroyed. But before his end, he activated a network of secondary resonators across

MarBryn. He claimed his followers would trigger them, spreading decay through the planetary beryllium veins. We recovered his mask; it may contain the key to locating those devices."

The councilwoman, Archmage Seraphina, took the wrapped mask with careful hands. "Voss was once our brightest alchemist. His fall was a tragedy. If his network remains active, we face a threat greater than any single adversary. We must act swiftly. Our best arcanologists will analyze the mask, but it may take days."

Kaelen yawned, stretching. "With respect, Archmage, I've already started. The mask emits a unique arcano-signature—a modulated decay frequency. If I can access the Spire's resonance chamber, I can amplify the signal and triangulate its matching sources. We could have a map of every active resonator by nightfall."

Seraphina regarded the tinkerer with new respect. "You are resourceful. Very well. The resonance chamber is at your disposal. But you will need a navigator. The resonators could be hidden anywhere—deep forests, volcanic calderas, even underwater. Our best cartographer is currently in the field, but her apprentice is here. Lyra!"

A young woman stepped from behind the council delegation. She was perhaps seventeen, with short-cropped auburn hair and a face smudged with ink. A brass compass hung from her belt, its needle spinning lazily, and she clutched a leather satchel stuffed with rolled maps and a bulky new-world magnetometer. Her eyes, bright with curiosity, darted from Berwyn to the sky-skiff and back.

"Lyra is a geomagnetic navigator," Seraphina explained. "She cannot wield magic, but she can read the planet's magnetic fields like a book. She has charted ley lines, mineral veins, and old-world ruins across MarBryn. If the resonators are tapping beryllium deposits, she will find them."

Lyra bowed hastily, nearly dropping her satchel. "It's an honor, Wizard Berwyn! I've studied your work on atomic bond reinforcement. The way you stabilized the old dam in Rivermarch using beryllium lattice theory? Brilliant! I mean—I'm ready to help. I've already mapped the major beryllium veins. If the resonators are placed along them, I can predict the most likely locations."

Berwyn smiled, her empathetic nature warming to the girl's enthusiasm. "Welcome aboard, Lyra. We'll need your expertise. Zephyr, Terron, Kaelen—let's get to the resonance chamber. We have a conspiracy to unravel."

The resonance chamber occupied the Spire's highest level, a circular room lined with tuning forks of varying sizes, each forged from different alloys. At its center, a holographic projector—new-world technology powered by captured lightning—displayed a three-dimensional map of MarBryn. Kaelen connected the copper mask to a series of arcano-sensors while Lyra unfurled her hand-drawn charts, cross-referencing beryllium deposits with known ley lines.

"The mask's signal is weak but distinct," Kaelen said, adjusting dials. "It's a carrier wave, like a radio transmission. I'm picking up three echo signatures—matching frequencies from different locations. They're faint, but they're there."

Lyra pointed to her charts. "The primary beryllium veins run through the Sunken Delta and the Emberpeak Caldera. If I overlay Kaelen's signal coordinates..." She tapped commands into the projector, and two red dots appeared on the map. "There! The delta, and the caldera. Both have old-world alchemical outposts. Voss's followers must have occupied them."

Terron rumbled, his voice echoing in the chamber. "The Sunken Delta contained beryllium for the Sunken Laboratory. If a resonator is there, it could destabilize the entire mountain range."

Zephyr crackled with agitation. "And the Emberpeak Caldera is an active volcano! If decay magic weakens the rock, it could trigger an eruption that buries half the province."

Berwyn studied the map, her strategic mind already forming a plan. "We can't split up—

we're stronger together. We'll need to disable each resonator in sequence. The Sunken Delta is the closest. We'll start there, then move to the caldera. Lyra, can you plot a route that minimizes travel time?"

Lyra beamed, pulling out a compass and a magnetometer. "Absolutely! The magnetic currents are favorable this season. If we ride the northern ley line, we can reach the Sunken Delta outpost in two days. I can navigate through the magnetic anomalies that usually throw off compasses.

Kaelen grinned. "I like her. She's got the right mix of brains and enthusiasm."

Archmage Seraphina placed a hand on Berwyn's shoulder. "The Council will provide supplies and reinforcements if needed. But be cautious—Voss's followers, they call themselves Energumens, will not surrender easily. They are fanatics, convinced that decay is the path to transcendence. They will fight to the last."

Berwyn nodded, her beryllium wand humming at her belt. "We've faced worse. We'll stop them, one resonator at a time. Lyra, welcome to the team. Gather your gear and meet us at the skiff in an hour."

As the group dispersed to prepare, Berwyn lingered at the map, tracing the red dots with a gloved finger. The Rust Alchemist was gone, but his shadow stretched across the land. She thought of the freed elementals, the corrupted constructs, and the fragile hope that still flickered in the hearts of her companions. Her empathetic nature sensed their determination, their trust in her leadership. She would not let them down.

An hour later, the Silver Zephyr lifted off from the Crystal Spire, its hull reinforced with a fresh beryllium force field. Berwyn stood at the helm, Kaelen beside her with the scanner, Zephyr perched on the bow, Terron anchored at the stern, and Lyra cross-legged on the deck, her maps spread before her. The wind carried them north, toward the Sunken Delta, where the first resonator awaited.

"We're on our way," Berwyn said, her voice steady. "Let's show Voss's followers that decay is not the only force in the universe. Bonds can be broken, but they can also be reforged—stronger than before."

The skiff sailed into the morning light, its passengers ready for the battles ahead.

Chapter 8: The Sunken Delta

The Silver Zephyr skimmed low over the Sunken Delta, a vast expanse of murky water, twisted cypress trees, and crumbling stone ruins half-submerged in the brackish tide. Once, this had been a thriving port city, its canals bustling with trade. Now, the old-world buildings leaned at drunken angles, their foundations eroded by centuries of neglect and the slow, inexorable creep of decay magic. New-world arcano-beacons, placed by the Elemental Council to warn travelers of unstable ground, flickered feebly in the mist.

Lyra consulted her magnetometer, her brow furrowed. "The resonator's signal is coming from the old customs house, right at the delta's heart. But the magnetic field here is all over the place—the water is saturated with dissolved beryllium salts from upstream mining. It's interfering with my readings."

Kaelen peered over the side of the skiff, her goggles splattered with moisture. "I see movement in the water. Something big. And it's not a fish."

Zephyr fluttered nervously. "I can scout from above, but the mist is thick. I might not see an ambush until it's too late."

Terron, standing at the bow, pointed a crystalline finger toward a cluster of ruins. "There. A dock that still stands. We can moor the skiff and proceed on foot—or by boat. The customs house is accessible only by water; the old bridges have collapsed."

Berwyn guided the skiff toward the dock, her beryllium senses reaching out. The water was alive with a low, discordant hum—the decay frequency, amplified by the dissolved beryllium. "The resonator is active, but it's not at full power yet. We have time. Kaelen, prepare the disruptor. Lyra, find us a boat. Terron, you're too heavy for most vessels; can you wade through the canals?"

The sentinel nodded. "The water is shallow here. I will follow alongside and guard against underwater threats."

They moored the skiff and located a rusted but serviceable flat-bottomed boat, its old-world motor long dead but its oars intact. Lyra took the oars, her navigator's instincts guiding them through the labyrinth of canals. Kaelen sat at the bow, disruptor ready, while Zephyr circled overhead. Berwyn stood in the center, wand glowing, her force field extended in a shimmering dome around the boat.

As they rounded a corner, a voice called out from the mist—a woman's voice, strained and desperate. "Help! Please, they've trapped me!"

Berwyn tensed, but her empathetic nature sensed genuine fear, not deception. "Where are you? Show yourself!"

A figure emerged from behind a collapsed pillar: a young woman with skin the color of sea foam and hair that flowed like liquid silver. Her lower body dissolved into a coil of water, marking her as a naiad—a water elemental. She was bound by chains of rusted iron that glowed faintly with decay magic, and her luminous eyes were wide with exhaustion.

"I am Nerida," she gasped. "I was the guardian of this delta, but I was captured by fanatics of the Rust Alchemist. They. . .the Energumens. They're using my connection to the water to amplify their resonator. If you free me, I can help you stop them."

Kaelen examined the chains with her scanner. "These are infused with the same decay frequency as the resonators. If I use the disruptor on them, it might weaken the bonds, but I'll need a few minutes."

Berwyn knelt beside the naiad. "Hold still. We'll get you out." She placed her wand against the chains, channeling beryllium magic not to break them, but to reinforce the healthy iron atoms while Kaelen's disruptor targeted the decay. Together, their powers worked in harmony—the chains first stabilized, then shattered into harmless rust.

Nerida slumped, then rose, her watery form strengthening as she touched the canal. "Thank you. The Energumens are in the customs house, three of them. They have a resonatorbut they've also summoned a guardian—a corrupted leviathan from the deep. It patrols the waters around the building."

Zephyr swooped down. "A leviathan? How big?"

"Large enough to swallow this boat whole," Nerida said grimly. "But I can calm the waters, create a current that will carry us past it. If we're quick and quiet."

Berwyn nodded. "Then let's move. Nerida, can you guide us through a back entrance?"

The naiad smiled, a hint of her old spirit returning. "I know every hidden canal and drainage tunnel. Follow me."

With Nerida swimming ahead, her form merging with the water, the boat glided silently through a narrow, overgrown channel. Terron waded behind, his crystalline body barely disturbing the surface. They emerged at the rear of the customs house, a once-grand building of marble and brass now stained with orange decay. A wooden dock led to a shattered window.

Inside, the resonator pulsed atop a stone altar, its cables snaking into the water below. Three Energumens stood guard, their rusted masks turning nervously. One held a curved sword; another, a crossbow; the third, a staff crackling with decay energy. And in the water beyond the altar, a massive shape coiled—the leviathan, its serpentine body covered in barnacles and glowing orange veins.

Berwyn raised her wand. "Kaelen, disable the resonator. The rest of us will handle the Energumens and the leviathan. Nerida, can you distract the creature?"

"I can do more than distract," Nerida said, her eyes flashing. She raised her arms, and the water around the leviathan churned, forming a whirlpool that pulled the beast away from the building. It roared, thrashing against the current.

The Energumens attacked. The swordsman charged Berwyn, his blade trailing rust. She met him with a beryllium-forged energy blade, the two weapons clashing in a shower of sparks. The crossbowman fired at Lyra, who ducked behind a pillar, then threw a handful of magnetic dust that jammed the weapon's mechanism. The staff-wielder sent a wave of decay at Zephyr, but the sky-drake dodged and retaliated with a lightning bolt that shattered the staff.

Terron lumbered through the wall, his crystalline fists smashing the resonator's support cables. "The device is vulnerable! Destroy it now!"

Kaelen sprinted to the altar, disruptor in hand. She slammed it against the control panel and reversed the polarity. The resonator screamed, its orange light flickering. But this time, the leviathan, sensing the disruption, broke free of Nerida's whirlpool and lunged at the building.

Berwyn saw it coming. She threw up the strongest force field she had ever conjured, a dome of pure beryllium energy that encased the entire customs house. The leviathan slammed into it, its decay-enhanced strength cracking the barrier. Berwyn gritted her teeth, feeling the strain in every atom. "Kaelen, hurry!"

"Almost... there!" Kaelen shouted. The resonator's hum died, and the orange light went dark. The leviathan convulsed, the decay veins on its body fading. It sank back into the depths, freed from the corruption.

The remaining Energumens, seeing their weapon destroyed and their guardian defeated, fled into the canals. Nerida let them go, her whirlpool dissipating.

Berwyn lowered her shield, exhausted but triumphant. "Two down. One to go."

Nerida swam to her side. "You saved my delta. I am in your debt. If you will have me, I would join you. The Emberpeak Caldera is a place of fire and steam—my water magic may be of use there."

Berwyn smiled, placing a hand on the naiad's shoulder. "Welcome aboard, Nerida. We'll need all the help we can get."

As the group regrouped on the dock, Lyra checked her magnetometer. "The caldera's resonator is the most powerful—it's sitting directly on a major beryllium vent. If it activates, the eruption could be catastrophic."

"Then we don't let it activate," Berwyn said. "To Emberpeak."

Chapter 9: The Heart of Emberpeak

The Emberpeak Caldera rose before them like a wound in the world, its rim jagged and blackened, its depths glowing with rivers of molten rock. The air shimmered with heat and the scent of sulfur, and the sky above was stained orange by the reflected fire. The Silver Zephyr circled once before Berwyn guided it to a precarious landing on a shelf of cooled lava. Terron stepped off first, his crystalline body refracting the volcanic glow into dancing rainbows, while Nerida coiled her watery form close to the skiff, visibly uncomfortable in the oppressive heat.

Lyra consulted her magnetometer, her face pale. "The resonator is directly below us, inside the caldera's throat. It's tapping the beryllium vent—the signal is stronger than anything we've seen. If it activates fully, it won't just cause an eruption. It could fracture the tectonic plate."

Kaelen wiped sweat from her brow, her disruptor already humming. "Then we shut it down fast. But how do we get down there? The heat alone will cook us."

Nerida stepped forward, her voice rippling. "I can create a sheath of water around us, a cooling barrier. But it will only last a few minutes in this heat. We must be swift."

Zephyr, perched on Berwyn's shoulder, crackled with static. "I'll fly ahead and scout. The

Energumens will be guarding the resonator. I can draw their fire while you descend."

Berwyn nodded, her amber eyes reflecting the lava's glow. "Terron, you'll lead the descent. Your body can withstand the heat. Kaelen, stay close to Nerida. Lyra, you'll guide us to the exact location. We move as one. No one gets left behind."

They advanced to the caldera's edge, where a narrow, winding path carved into the rock descended into the inferno. Nerida raised her arms, and a shimmering cocoon of water enveloped the group, hissing as it met the superheated air. They began the descent, Terron's heavy footsteps crunching on volcanic glass, Zephyr darting ahead into the smoke.

The path opened into a vast cavern, its ceiling lost in shadow, its floor a lake of bubbling magma. At the center, on an island of cooled obsidian, stood the final resonator—a monstrous device of black iron and pulsating orange crystal, its cables plunging into the lava to draw power from the beryllium vent. Around it, a dozen Energumens in rusted robes chanted, their voices a discordant hymn. And before them, wielding a staff of twisted beryllium and decay, stood their leader—a woman in a mask of corroded silver, her eyes burning with fanatical zeal.

"The beryllium witch!" she shrieked. "You are too late! The Resonator is at full charge! In moments, the decay will spread through the planetary veins, and MarBryn will crumble!"

Berwyn stepped forward, her wand blazing. "I've heard that before. Your master said the same thing. He's dust now. You will be too, unless you stand down."

The leader laughed, raising her staff. "I am the Rust Prophetess! I carry the Master's will! Attack!"

The Energumens surged forward, weapons raised—swords, crossbows, and staves crackling with decay. Zephyr dove from above, unleashing a lightning storm that sent several sprawling. Terron charged, his crystalline fists shattering weapons and armor. Nerida swept a wave of water across the obsidian, knocking Energumens off their feet and cooling the ground enough for Kaelen and Lyra to advance.

Berwyn engaged the Prophetess directly, their wands clashing in a shower of green and orange sparks. "You're a fool," Berwyn said, her voice steady despite the strain. "Decay isn't transcendence. It's just destruction. There's no renewal in what you're doing—only oblivion."

"You understand nothing!" the Prophetess snarled, pushing harder. "The old world was built on bonds that always break! We are freeing MarBryn from the illusion of permanence!"

Berwyn felt the woman's despair, a hollow echo of Voss's madness. She channeled her beryllium magic not into attack, but into a wave of empathy, projecting her own certainty—the strength of bonds, the beauty of connection. "You're wrong. Bonds can be reforged. They can be stronger than before. That's what beryllium teaches us. It's light, but it's rigid. It supports without weighing down. That's what we can be—for each other."

The Prophetess faltered, her staff wavering. In that moment, Kaelen reached the resonator, disruptor in hand. "Cover me!" she shouted.

Lyra threw a handful of magnetic dust at the control panel, causing the Energumens' weapons to spark and malfunction. Nerida doused the chanting Energumens with a wave of water, breaking their concentration. Zephyr and Terron held the line, keeping the remaining enemies at bay.

Kaelen slammed the disruptor against the resonator's core. "Reversing polarity!"

The device screamed, its orange light flickering. But the Prophetess, recovering, lunged at Kaelen with her staff. Berwyn intercepted, lightening her body and leaping between them. She caught the staff with her wand, and with a surge of beryllium magic, she forced the bonds in the staff to become perfectly rigid—then brittle. The staff shattered.

The Prophetess stumbled back, her mask cracking. "No... the Master..."

"The Master is gone," Berwyn said. "It's over."

The resonator's hum died. The orange light went dark. The cables fell slack, and the crystal core crumbled into inert dust. The ground trembled, but instead of an eruption, the magma began to cool, the beryllium vent stabilizing as the decay frequency vanished.

The remaining Energumens, seeing their leader defeated and their weapon destroyed, threw down their arms. The Prophetess sank to her knees, her mask falling away to reveal a woman's face, tear-streaked and exhausted. "I... I only wanted to end the pain. The decay... it was all I had left."

Berwyn knelt beside her, her voice gentle. "There's another way. It starts with rebuilding—not destroying. We can help you find it."

Terron approached, his molten eyes soft. "The caldera is stable. The vent is sealed. The network is broken."

Lyra checked her magnetometer, a grin spreading across her face. "No more signals. The resonators are offline. MarBryn is safe."

Zephyr landed on Berwyn's shoulder, his scales dim but triumphant. "We did it. We actually did it."

Kaelen slumped against the resonator, laughing with relief. "I need a vacation. A long one. Somewhere with no rust, no Energumens, and definitely no volcanoes."

Nerida smiled, her watery form rippling. "The delta will bloom again. The rivers will run clean. You have saved us all, Berwyn."

Berwyn stood, looking at her companions—her friends. Kaelen, the brilliant tinkerer who had never given up. Zephyr, the loyal sky-drake with a spark of lightning in his heart. Terron, the ancient sentinel who had found redemption. Lyra, the young navigator whose courage matched her skill. Nerida, the naiad who had reclaimed her home. And even the Prophetess, a broken woman who might yet heal.

"Let's go home," Berwyn said. "All of us."

The journey back to Lumina was quiet, the Silver Zephyr sailing through clear skies. When they landed at the Crystal Spire, the Elemental Council greeted them with cheers and tears. The freed elementals—Embera among them—had spread the word of Berwyn's deeds. The Energumens were taken into custody, offered rehabilitation instead of punishment. The Prophetess, whose true name was Elara, began the long road to recovery under the Council's guidance.

In the weeks that followed, Berwyn oversaw the restoration of the damaged regions. The Sunken Delta's waters ran clear, and Nerida returned to her guardianship. Terron chose to stay in Lumina, becoming a protector of the Spire. Lyra was promoted to chief cartographer, her maps now including the safe beryllium veins. Zephyr remained Berwyn's constant companion, his static a comforting hum. And Kaelen, after a well-deserved rest, began work on a new invention—a device that could detect and neutralize decay magic before it took root.

Berwyn herself returned to her studies, but with a new perspective. She wrote extensively on the dual nature of beryllium—its strength and its potential for brittleness, its ability to build and its susceptibility to corruption. Her treatise, "The Bonds That Hold Us", became a cornerstone of alchemical education.

One evening, as the sun set over the Crystal Spire, Berwyn stood on her balcony, her beryllium wand glowing softly. She thought of the journey—the battles, the losses, the triumphs. She thought of Voss, a cautionary tale of genius twisted by despair. She thought of her friends, whose bonds had been tested and proven unbreakable.

Zephyr landed beside her. "What are you thinking about?"

Berwyn smiled, her amber eyes reflecting the stars. "About how even the lightest element can be the strongest. About how bonds—whether between atoms or between people—are what hold the world together. And about how, as long as we have each other, no decay can ever truly win."

The sky-drake nuzzled her cheek. "That's very philosophical. Also, Kaelen says dinner's ready. She made something she calls 'alchemical stew.' I'm a little worried."

Berwyn laughed, the sound bright and free. "Then let's not keep her waiting."

Together, they walked inside, leaving the balcony and the stars behind, but carrying the light of their adventure forward into a future full of promise. The latest adventure of Berwyn the Alchemical Wizard had ended, but the bonds she had forged would last forever.

Beryllium Facts

- First discovered in 1798 in France
- Steel-gray hard metal
- Good conductor of heat and electricity
- Non-magnetic
- Not ductile, malleable when warm
- Alkaline Earth Metal

Note From The Author

Thank you for stepping into MarBryn with Berwyn and her companions. This book grew from two quiet convictions I carry into every story: that curiosity is a superpower, and that even the hardest challenges are solvable when we lean on others.

First, I wanted to honor the wonder of real-world science. Berwyn's magic is built from the properties of a real element—beryllium—so readers can sense that the fantastical is anchored in something genuine. Chemistry isn't just a backdrop here; it's the engine of the plot, the rules of the world, and the source of both beauty and peril. If something in this story sparks recognition, I hope it invites you to pause, consider, and see the magic in everyday knowledge.

Second, this story is really about bonds. It's about turning strangers into a team, fear into courage, and decay into renewal. Berwyn's leadership is quiet and inclusive, grounded in empathy as much as power. The companions—Kaelen's ingenuity, Zephyr's courage, Terron's memory and endurance, Lyra's curiosity, Nerida's gentleness, and Embera's fire—each contribute a thread to the tapestry. My aim was to show that the strongest force in any world (whether made of atoms or people) is the trust we build together.

The world of MarBryn blends old-world machines with new-world magic, a deliberate homage to the ways history, tradition, and invention inform one another. I pictured landscapes that spark the senses—the glass desert's prisms, the echoing halls of the Sunken Laboratory, the heat and dawn of

Emberpeak—because settings should do more than frame action: they should breathe and sing.

If you've found joy here, I hope you'll read about the other elements in the series as they move toward new challenges and new friends. And if you've thought about a science lesson or a chemistry curiosity while turning these pages, that's exactly the spark I chase: to make learning feel like an adventure worth pursuing.

With gratitude for your imagination and your curiosity,

Sybrina Durant

PS: If you're interested in reading the resources the scientific data in this book was derived from, check out this webpage: ***https://magicalptelements.com/mw2pt/***

About The Author

Sybrina Durant is the author of

"The Magical Elementals Chemistry Fantasy Series"

"The Magical Elements of the Periodic Table Series"

"The No Metal No Magic Series"

"The Blue Unicorn - Journey To Osm" books

and she is the creator of the "Elemental Fantasy Periodic Table" which is available as a brilliantly colored poster.

There are also book activities, songs, trading cards and t-shirts for all 118 elements and elementals, too.

Visit MagicalPTElements.com to see everything that is currently available.

If you love to sing and learn, check out the "Chemistry Sing Alongs" playlist on the Sybrinad channel on Youtube.

Glossary of Scientific Terms in Berwyn the Alchemical Wizard with Beryllium Magic

Alloys - Strong, durable metallic mixtures created by blending a metal like beryllium with other metals. Berwyn's world runs on them: the Silver Zephyr's hull is a beryllium-aluminum alloy, and the old masters forged beryllium alloys deep in the Foundry of Forgotten Alloys — proof that Berwyn's magic works hand in hand with real metallurgy rather than replacing it.

Alchemical Motes - Floating, microscopic particles of magical energy released during alchemical processes. They swirl from Berwyn's wand — golden and scented of ozone — every time she channels beryllium magic, a visible signature of the atomic bonds she's weaving or strengthening in that moment.

Alchemical Wizardry / Alchemy - A discipline blending science and magic focused on the study and manipulation of matter, chemical properties, and atomic bonds. It is Berwyn's entire craft: every spell she casts — shields, lattices, force fields — is really an act of sensing and reshaping atomic bonds through beryllium.

Alkaline Earth Metal - The specific chemical group and family of elements to which beryllium belongs. This family membership explains the paradox at the heart of Berwyn's power — beryllium is light enough to lift a sky-skiff, yet rigid enough to stop a leviathan — and why her magic can be both a shield and, in the wrong hands, a weapon.

Atomic Bonds - The microscopic physical and chemical forces that connect atoms together to form all matter in the world. Berwyn's entire magic system is built on sensing and reinforcing these bonds — her force fields are literally 'woven from strengthened atomic bonds' — while the Rust Alchemist's decay magic exists to unravel them. The whole story is this contest, bond against bond.

Beryllium - Element 4 - a light, strong, steel-gray alkaline earth metal known for its rigidity and high strength. It is the literal source of Berwyn's magic — her wand is tipped with a shard of pure beryllium, her gloves are stitched with beryllium thread, and every shield, lattice, and blade she conjures channels this one element.

Beryllium Force Field - A protective barrier created by reinforcing and weaving together atomic bonds. This is Berwyn's signature spell, used again and again: shielding the Silver Zephyr from silica-wraiths, encasing the caged elementals in the Transmutation Chamber, and holding back a corrupted leviathan at the Sunken Delta.

Beryllium-Aluminum Alloy - A lightweight, highly durable metal mixture used to build resilient structures like sky-skiff hulls. The Wind-Sisters forge the Silver Zephyr from it, and Kaelen marvels at its off-the-charts tensile strength — the same real-world property Berwyn later amplifies with her own beryllium magic to armor the ship against wraith attacks.

Beryllium Lattice - A highly structured, rigid molecular formation of beryllium atoms. Berwyn invokes this exact structure in combat — Lyra even praises her for stabilizing Rivermarch's old dam using 'beryllium lattice theory' — locking a silica-wraith's body or the Ferroclast's iron into perfect rigidity until it shatters under its own weight.

Beryllium Vent - A natural subterranean fissure rich in concentrated beryllium deposits. The Rust Alchemist's final Resonator is built directly over one inside the Emberpeak Caldera, tapping the vent to try to trigger a decay chain reaction — the same beryllium source Berwyn's magic depends on, turned into humanity's undoing.

Covalent Bonds - Strong chemical bonds formed when atoms share electrons, granting structural stability. Berwyn explains that beryllium's strength comes from forming these bonds through its electron configuration, and she channels that same covalent strength to reinforce Terron's silica matrix and the Ferroclast's failing iron structure.

Decay Magic / Corruption - Dark alchemical magic that targets and unravels atomic bonds, causing materials to rust and crumble. It is the Rust Alchemist's power and the direct opposite of everything Berwyn's beryllium magic stands for — where she reinforces bonds, his magic breaks them, and the entire plot is the race to stop his corruption from spreading through MarBryn's beryllium veins.

Decay Frequency - A specific wave vibration designed to break down atomic connections. Voss's Resonators broadcast this frequency to unravel matter at range; Berwyn and Kaelen must counter it by tuning the resonance disruptor and Berwyn's own beryllium resonance against it in every showdown.

Electron Configuration - The specific arrangement of electrons around an atom's nucleus that dictates how it bonds with other elements. Berwyn cites beryllium's electron configuration directly as the reason it forms such strong, rigid covalent bonds — the scientific principle underlying every shield and lattice she casts.

Geomagnetic Navigation - A method of charting terrain and finding locations by reading planetary magnetic fields. It is Lyra's non-magical skill, and it becomes essential to Berwyn's quest: without magic of her own, Lyra reads MarBryn's magnetic fields to trace the beryllium veins the Rust Alchemist's followers are exploiting and pinpoint each hidden Resonator.

Kinetic Energy - Energy possessed by an object due to its motion, which certain constructs can absorb. The Ferroclast feeds on kinetic energy, growing stronger with every blow it takes — which is exactly why Berwyn defeats it not with force, but by using beryllium magic to lock its iron atoms into a lattice so rigid it shatters under its own weight.

Ley Lines - Natural channels of magnetic and magical energy crisscrossing the planet. Lyra follows them to plot the fastest route between resonators, tying the beryllium deposits that fuel both Berwyn's magic and Voss's weapon into the wider magical geography of MarBryn.

Magnetometer - A specialized device used to measure the strength and direction of magnetic fields. Lyra's magnetometer is how the team locates every beryllium-tapping Resonator across MarBryn — from the Sunken Delta's dissolved beryllium salts to the vent beneath Emberpeak — turning Berwyn's element-based magic into something trackable with real instruments.

Molecular Structure - The 3D physical arrangement of bound atoms that makes up a substance. Berwyn reinforces her own molecular structure when she duels Voss directly, using beryllium magic defensively even on her own body, while his decay magic tries to unweave that same structure atom by atom.

Ozone - A distinct-smelling gas formed during high-energy electrical and magical discharges. It accompanies the alchemical motes swirling from Berwyn's wand and the crackle of Zephyr's lightning alike, grounding her high-energy beryllium spellcasting in a real chemical byproduct.

Quartz / Beryllium Prisms - Crystalline tools used to bend, refract, and isolate light into distinct frequencies. Berwyn is studying one in the opening chapter, noting how each facet bends light into a different frequency — introducing the very idea of resonant frequency that later becomes the battleground between her beryllium magic and Voss's decay-frequency Resonators.

Resonance Disruptor - A device emitting counter-vibrations to stabilize or alter target atomic frequencies. Kaelen builds and refines it throughout the story specifically to counter decay frequency; paired with Berwyn's beryllium magic, it becomes the one-two punch that disables every Resonator the team faces.

Resonator - An arcano-chemical machine used to amplify and transmit atomic decay waves. Voss builds a network of them, each one drawing on a beryllium vein or vent — the same element that powers Berwyn's magic — to try to trigger a world-ending chain reaction, making the Resonator the direct antagonist to her beryllium force fields.

Silica Bonds - The molecular connections holding glass and silicon dioxide structures together. Berwyn's beryllium magic reaches into these bonds to freeze a silica-wraith's body rigid mid-attack, and later to reinforce — rather than break — Terron's corrupted silica matrix, which is how she frees him from the Rust Alchemist's control.

Tensile Strength - The maximum amount of stretching stress a material can withstand without snapping. Kaelen marvels at the beryllium-aluminum hull's off-the-charts tensile strength, the real-world property that explains why Berwyn can weave beryllium into shields and lattices strong enough to stop weapons, wraiths, and even a leviathan.

Thermal Shock - Rapid temperature changes that strain and weaken molecular bonds within a material. Zephyr's lightning and Embera's flame subject the Ferroclast's iron to thermal shock, straining its bonds — the weakness Berwyn then exploits with beryllium magic to lock those already-stressed atoms into a lattice so rigid it shatters. Transmutation: The alchemical process of altering one substance or element into another. It is the founding purpose of the Sunken Laboratory and the ambition behind Voss's Beryllium Heart — and its philosophical opposite is Berwyn's own magic, which reinforces and stabilizes bonds rather than transmuting or destroying them.

Arcano-Reactor - A self-contained power source blending arcane and mechanical engineering. It drives the Transmutation Chamber's Resonator; Kaelen reverses its polarity to create a feedback loop that shuts the decay weapon down, working in tandem with Berwyn's beryllium force field holding the elementals safe.

Beryllium Veins - The natural underground deposits of beryllium running throughout MarBryn, distinct from a concentrated vent. They are the very thing Berwyn's magic draws its power from and the thing Voss's Resonators hunt out — Lyra spends the second half of the story mapping them to predict where the next attack on Berwyn's element will strike.

Chain Reaction - A self-sustaining sequence in which each reaction triggers the next. Voss's ultimate plan inverts beryllium's stabilizing bonds so that decay propagates through every trace of beryllium on MarBryn in a chain reaction — the doomsday scenario Berwyn spends the entire book racing to prevent.

Feedback Loop - A process where a system's output is fed back into itself, amplifying or canceling its own signal. Kaelen repeatedly triggers feedback loops in Voss's machinery by reversing polarity on the resonance disruptor, the engineering counterpart to the beryllium magic Berwyn uses to hold the line while the loop takes effect.

Memory Alloy - A metal alloy engineered to 'remember' and return to a set shape or structure. The Ferroclast's body is a corrupted old-world memory alloy; Kaelen salvages a sample after Berwyn shatters it with beryllium magic, hoping to reverse-engineer a countermeasure against future decay-twisted constructs.

Silica Matrix - The interconnected molecular framework that gives a silica-based structure its shape and strength. It describes Terron's crystalline body; Berwyn reaches into his silica matrix with beryllium magic to reinforce his original, healthy bonds and push out the Rust Alchemist's corruption, freeing him.

www.ingramcontent.com/pod-product-compliance
Lightning Source LLC
LaVergne TN
LVHW050747061126
851857LV00013B/1032
9781942740735